PRESENTED TO

_____

FROM

_____

*Write your blessings,*
*Write down every one,*
*Write down every blessing,*
*Share what God has done.*

# Deliverance Diary
## Write Down Every Blessing

E.J.H. MOFFETT

# DELIVERANCE DIARY
# WRITE DOWN EVERY BLESSING

*Scripture taken from the New King James Version ®. Copyright © 1982 by Thomas Nelson, Inc. Used by permission. All rights reserved. Exceptions are noted as follows: English Standard Version (ESV), King James Version (KJV), New International Version (NIV), New Life Version (NLV), New Living Translation (NLT) and The Message (MSG).*

*iUniverse books may be ordered through booksellers or by contacting:*

*iUniverse*
*1663 Liberty Drive*
*Bloomington, IN 47403*
*www.iuniverse.com*
*1-800-Authors (1-800-288-4677)*

*ISBN: 978-1-4917-4991-3 (sc)*
*ISBN: 978-1-4917-4992-0 (e)*

*Printed in the United States of America.*

*iUniverse rev. date: 10/31/2014*

This first edition Deliverance Diary was created for you because of my children and grandchildren; in respect of faithful ones born May 11 and August 7; and because of my journey.

*a few flowers*™
25 August 2014
7:11 pm

Dear Hearts, you have 122 pages to create your first handwritten Deliverance Diary.

A diary is usually dated, lockable and very personal. A Deliverance Diary is different. It's special and sovereign. It is dated but not locked. Its pages are filled with special deliverances from God. Deliverances you will not be able to keep to yourself.

Extraordinary examples of deliverances to the faithful are written about in the eleventh chapter of Hebrews. There is also a spiritual definition of faith written in this chapter. Hebrews 11:1 tells us that "faith is the substance of things hoped for, the evidence of things not seen".

Did you know God blesses you according to the power of your own faith? *Ephesians 3:19-21*

You are going to rediscover the blessings God give to you in every way. You will not be able to keep this to yourself. God consistently and without any effort blesses us each day. We consistently and without any thought forget to count our blessings each day.

Listen again to Jesus who went up into a mountain, sat before His disciples, and taught the multitudes about blessings. They are still relevant and available to the faithful to this day.

"God blesses those who are poor
and realize their need for Him,
*for the Kingdom of Heaven is theirs.*
God blesses those who mourn,
*for they will be comforted.*

God blesses those who are humble,
*for they will inherit the whole earth.*
God blesses those who hunger and thirst for justice,
*for they will be satisfied.*
God blesses those who are merciful,
*for they will be shown mercy.*
God blesses those whose hearts are pure,
*for they will see God.*
God blesses those who work for peace,
*for they will be called the children of God.*
God blesses those who are persecuted for doing right,
*for the Kingdom of Heaven is theirs."*
*Meditate on Matthew 5:3-12*

The deliverances you write about will be like salt (study Ephesians 5:13); shine like a light (meditate on Ephesians 5:14a); and cannot be hidden (apply Ephesians 5:14b).

You will want to share your blessing with those you love, those you know, and those you meet new each day. We at *a few flowers*™ will be honored and blessed to meet you through your Deliverance Diary. Email one or many of your blessings to amen@afewflowers.com. Share what God has done. Begin my friend.

GIVE ALL YOUR ATTENTION
TO WHAT GOD IS DOING FOR YOU RIGHT NOW.
*Matthews 6:34 MSG*

DELIVERANCE DATE | _____

DON'T EVEN THINK ABOUT IT—
TOMORROW, THAT IS.
*Matthews 6:34 NLV*

DELIVERANCE DATE | _____

SEEK THE KINGDOM OF GOD ABOVE ALL ELSE,
AND LIVE RIGHTEOUSLY,
AND HE WILL GIVE YOU EVERYTHING YOU NEED.
*Matthews 6:33*

DELIVERANCE DATE | _____

LAUGH AT YOUR TROUBLES.
WATCH THEM SHRINK BECAUSE OF YOUR FAITH.
*Matthews 6:30*

DELIVERANCE DATE | _____

Seek God's kingdom and righteousness,
and "things" will be added to your life.

*Matthews 6:33 KJV*

Deliverance Date | _____

WATCH. STAND FAST IN THE FAITH.
BE BRAVE. BE STRONG.

*1 Corinthians 16:33*

DELIVERANCE DATE | _____

AND WE KNOW THAT ALL THINGS WORK TOGETHER
FOR GOOD TO THEM THAT LOVE GOD,
TO THEM WHO ARE THE CALLED
ACCORDING TO HIS PURPOSE.

*Romans 8:28 KJV*

DELIVERANCE DATE | _____

HE WILL SHOW YOU THE PATH OF LIFE.

*Meditate on Psalm 16:11a*

DELIVERANCE DATE | _____

In His presence you'll experience joy.

*Meditate on Psalm 16:11b*

Deliverance Date | _____

AT HIS RIGHT HAND ARE PLEASURES FOREVERMORE.

*Meditate on Psalm 16:11c*

DELIVERANCE DATE | _____

REFUSE BITTERNESS.
*Ephesians 4:31*

DELIVERANCE DATE | _____

ANTICIPATE PEACE.
*Philippians 4:7*

DELIVERANCE DATE | _____

WHEN YOU EXPERIENCE HOPE WITH JOY
YOU ARE IN THE DEEPEST PART OF PEACE.
*John 14:26-28*

DELIVERANCE DATE | _____

Your glass is not half empty.
Look at it again.
It certainly is half full.

*1 Corinthians 4:7, 8*

Deliverance Date | _____

WAR DOES NOT END QUIETLY.
GIVE PEACE ITS CHANCE.

*Exodus 14:14*

DELIVERANCE DATE | _____

Let no one reposition you
so you cannot see the omniscient Jesus.
No one else has infinite awareness,
understanding, and insight about you.

*Romans 8:34-36*

Deliverance Date | _____

LET NOTHING SEPARATE YOU
FROM THE OMNIPRESENT JESUS.
NO ONE ELSE CAN BE PRESENT WITH YOU
IN ALL PLACES AT ALL TIMES.
*Romans 8:34-36*

DELIVERANCE DATE | _____

LET NOTHING DISLOCATE YOU
SO YOU CANNOT EXPERIENCE THE
OMNIPOTENT JESUS. NO ONE ELSE
HAS COMPLETE AND UNLIMITED POWER TO GIVE YOU.

*Romans 8:34-36*

DELIVERANCE DATE | _____

HISTORY CONTAINS VIOLENCE.
SHARPEN YOUR PEACEMAKING SKILLS
NOT YOUR WEAPONRY.
*Matthews 5:8-10*

DELIVERANCE DATE | _____

WHEN YOU PRAY ASK GOD TO
MAKE YOU MORE WORTHY,
MORE OF A SERVANT AND MORE KNOWING.
*Matthew 10:12-14*

DELIVERANCE DATE | _____

WHEN YOU MEDITATE ASK GOD FOR WISDOM.

*James 1:5*

DELIVERANCE DATE | _____

INCLINE YOUR EAR TO **WISDOM**,
APPLY YOUR HEART TO **UNDERSTANDING**.

*Proverbs 2:1-6*

DELIVERANCE DATE | _____

Say to **wisdom**, "You *ARE* my sister,"
And call **understanding** "*YOUR* nearest kin."

*Proverbs 7:4*

Deliverance Date | _____

THE LORD BY **WISDOM** FOUNDED THE EARTH;
BY **UNDERSTANDING** HE ESTABLISHED THE HEAVENS.

*Proverbs 3:19*

DELIVERANCE DATE | _____

THE FEAR OF THE LORD
*IS* THE BEGINNING OF **WISDOM,**
AND THE KNOWLEDGE OF THE HOLY ONE
*IS* **UNDERSTANDING.**

*Proverbs 9:10*

DELIVERANCE DATE | _____

> WISDOM *IS* THE PRINCIPAL THING.
> *THEREFORE* **GET** WISDOM.
> AND IN ALL YOUR **GETTING**,
> **GET UNDERSTANDING.**
> *Proverbs 4:7*

DELIVERANCE DATE | _____

DELIVERANCE DATE | _____

When God first sent His Holy Spirit
it felt like a rushing might wind.
*Acts 2:1, 2*

Deliverance Date | _____

ARE YOU CURIOUS ABOUT ANGELS?
*Hebrews 13:2*

DELIVERANCE DATE | _____

GOD COMMANDED AND THE ANGELS WERE CREATED.
*Psalm 148:2-4*

DELIVERANCE DATE | _____

ANGELS ARE THE SONS OF GOD.
THEY SHOUTED FOR JOY DURING THE CREATION.
*THEY WERE THERE YOU KNOW.*
*Job 38:4-7*

DELIVERANCE DATE | _____

ANGELS ARE SPIRITS BUT NOT GOD'S BEGOTTEN SONS.
*Hebrews 1:5-7*

DELIVERANCE DATE | _____

ANGELS ARE MESSENGERS WHO GUARDED
AND RESCUED GOD'S PEOPLE IN APOSTOLIC TIMES.
*Daniel 6:20, 21; Acts 5:17-21*

DELIVERANCE DATE | _____

THE FIRST TIME ANGELS APPEAR IN
SCRIPTURE IS GENESIS THREE
WHEN ADAM AND EVE WERE SENT
OUT OF THE GARDEN OF EDEN.
THEY BARRED THE ENTRYWAY.
*Genesis 3:17-24*

DELIVERANCE DATE | _____

THE SECOND TIME ANGELS APPEAR IN SCRIPTURE
IS GENESIS EIGHTEEN WHEN ABRAHAM AND SARAH
WERE TOLD GOD WOULD GIVE THEM
A SON IN THEIR OLD AGE.

*Genesis 18:1-15*


DELIVERANCE DATE | _____

ANGELS USUALLY APPEAR AS MEN
WEARING SHINY WHITE GARMENTS
RATHER THAN WOMEN OR CHILDREN;
OR WITH HARPS, WINGS OR HALOS.
*Luke 24:1-8*

DELIVERANCE DATE | _____

THERE ARE 300 REFERENCES TO ANGELS IN THE BIBLE.
ONLY THE ARCHANGELS MICHAEL
AND GABRIEL ARE CALLED BY NAME.

*Jude 1:8-10; Luke 1:18-20*

DELIVERANCE DATE | _____

Angels did supernatural things
during Biblical times.

*1 Corinthians 18:8-10*

Deliverance Date | _____

ANGELS ARE LIKE SECRET AGENTS.
THEY CARRY GOD'S PROVIDENCE.

*Genesis 24:7*

DELIVERANCE DATE | _____

GOD MADE AN INNUMERABLE COMPANY OF ANGELS.

*Hebrews 12:22*


DELIVERANCE DATE | _____

ANGELS WERE MADE TO WATCH OVER US
BOTH DAY AND NIGHT.

*Luke 4:9*

DELIVERANCE DATE | _____

WE ARE NOT ANGELS.
*Psalm 8:3-5*

DELIVERANCE DATE | _____

GOD MADE US, MANKIND,
A LITTLE LOWER THAN THE ANGELS.
*Psalm 8:5*

DELIVERANCE DATE | _____

WE CAN ONLY BECOME <u>LIKE</u> ANGELS
BUT ONLY AFTER WE DIE.
*Luke 20:34-36*

DELIVERANCE DATE | _____

JESUS IS NOT AN ANGEL
IN FACT, HE'S BETTER THAN THEM.
*Hebrews 1:3-5*

DELIVERANCE DATE | _____

(Our) God *is* Spirit, and those who worship Him
must worship in spirit and truth.

*John 4:24*

Deliverance Date | _____

REFUSE UNTRUTHFULNESS.

*Genesis 3:4*

DELIVERANCE DATE | _____

WELCOME TRUTH.
*John 8:31, 32*

DELIVERANCE DATE | _____

GOD GIVES YOU NEW MERCIES EVERY MORNING.
*Lamentations 3:22, 23*

DELIVERANCE DATE | _____

IN GOD'S WAY MANY OF OUR GREATEST GAINS
COME AFTER OUR LOSSES.

*Philippians 3:7*

DELIVERANCE DATE | _____

...COUNT ALL THINGS LOSS FOR THE EXCELLENCE
OF THE KNOWLEDGE OF CHRIST JESUS (YOUR) LORD...
*Philippians 3:8a*

DELIVERANCE DATE | _____

SUFFER THE LOSS OF ALL THINGS,
AND COUNT THEM AS RUBBISH AND GAIN CHRIST.
*Philippians 3:8b*

DELIVERANCE DATE | _____

Be found in Him, not having
(your) own righteousness,
which *is* from the law, but that which
*is* through faith in Christ.
the righteous, which is from God by faith.
*Philippians 3:9*

Deliverance Date | _____

FORGET THOSE THINGS WHICH ARE BEHIND
AND REACH FORWARD TO THOSE
THINGS WHICH ARE AHEAD.
*Philippians 3:12-16*

DELIVERANCE DATE | _____

YOUR REAL CITIZENSHIP IS IN HEAVEN
*Philippians 3:17-21*


DELIVERANCE DATE | _____

You must be saved to be a citizen in heaven.

*Acts 4:10-12*

Deliverance Date | _____

WHAT MUST YOU DO TO BE SAVED?
*Acts 4:10-12*

DELIVERANCE DATE | _____

THIS IS JESUS SPEAKING.
"I AM THE WAY, THE TRUTH, AND THE LIFE.
NO ONE COMES TO THE FATHER EXCEPT THROUGH ME".
*John 14:6*

DELIVERANCE DATE | _____

How do you come through Jesus
to the Father who is in Heaven
where your real citizenship lies?
*John 14:6*

Deliverance Date | _____

First, you must hear the gospel, understand it, and accept it. If you are not obedient to the gospel of Jesus you are lost. There is salvation in no other name.

*Romans 3:23; 2 Thessalonians 1:8; Acts 4:12*

Deliverance Date | _____

SECOND, YOU MUST BELIEVE AND HAVE FAITH.
WITHOUT FAITH IT'S IMPOSSIBLE TO PLEASE GOD.
*Hebrews 11:6*

DELIVERANCE DATE | _____

THIRD, YOU MUST REPENT AND BE CONVERTED
SO YOUR PAST SINS CAN BE BLOTTED OUT FOREVER.

*Acts 3:19*

DELIVERANCE DATE | _____

FOURTH, YOU MUST MAKE A CONFESSION.
SAY, "JESUS IS THE SON OF GOD".

*Romans 10:9, 10*

DELIVERANCE DATE | _____

FIFTH, YOU MUST BE BAPTIZED IN THE NAME OF JESUS CHRIST FOR THE REMISSION OF YOUR SINS; AND YOU SHALL RECEIVE THE GIFT OF THE HOLY SPIRIT.

*Acts 2:38*

DELIVERANCE DATE | _____

FINALLY, GOD ADDS YOU TO HIS CHURCH,
THE ONE YOU CAN READ ABOUT IN
YOUR COPY OF THE BIBLE.
HE ADDS YOUR NAME TO THE BOOK OF LIFE.
YOU MUST REMAIN FAITHFUL TO HIM UNTIL DEATH.
*Acts 2:47; Philippians 4:3*

DELIVERANCE DATE | _____

Cast your burden on the Lord,
and He shall sustain you.
He shall never permit
the righteous to be moved.

*Acts 17:30*

Deliverance Date | _____

REFUSE IGNORANCE.

*Acts 17:30*

DELIVERANCE DATE | _____

ANTICIPATE GODLY WISDOM.
*2 Corinthians 1:12*

DELIVERANCE DATE | _____

ARE YOU GIVING LOVE EACH DAY?
ARE YOU AUGUST-LIKE?
*John 13:33-35*

DELIVERANCE DATE | _____

LOVE UNCONDITIONALLY WHEN YOU LOVE.
YOU MAY NOT GET ANOTHER OPPORTUNITY
TO LOVE IN THIS WAY.
*1 Corinthians 13:1-e*

DELIVERANCE DATE | _____

ARE YOU ON THE RIGHTEOUS SIDE OF LIFE?
ARE YOU TAURUS-LIKE?

*Genesis 7:1-3*

DELIVERANCE DATE | _____

THE RIGHTEOUS SHALL INHERIT
THE LAND AND DWELL IN IT FOREVER.
THE MOUTH OF THE RIGHTEOUS SPEAKS WISDOM
AND HIS TONGUE TALKS OF JUSTICE.
*Psalm 37:29, 30*

DELIVERANCE DATE | _____

It's as good a day as any
to stop and smell the roses in your path.
This day will not come to you again.

*Psalm 2:1*

Deliverance Date | _____

REFUSE LONELINESS.

*Proverbs 9:2*

DELIVERANCE DATE | _____

ANTICIPATE JOY.
*Psalm 30:5*

DELIVERANCE DATE | _____

A SERVANT OF THE LORD MUST NOT QUARREL
BUT BE GENTLE TO ALL, ABLE TO
TEACH AND BE PATIENT.
*2 Timothy 2:24*

DELIVERANCE DATE | _____

REFUSE CONFUSION.
*1 Corinthians 14:33*

DELIVERANCE DATE | _____

ANTICIPATE HAPPINESS.
*Proverbs 10:12*

DELIVERANCE DATE | _____

.

DELIGHT YOURSELF IN THE LORD,
AND HE WILL GIVE YOU THE DESIRES OF YOUR HEART.

*Psalm 37:4*

DELIVERANCE DATE | _____

O taste and see that the Lord is good:
blessed is the man that trusteth in Him.

*Psalm 34:8 KJV*

Deliverance Date | _____

TAKE THE TOOTHPASTE TEST.
*JUST GOOGLE IT.*
*1 Corinthians 2:15, 16*

DELIVERANCE DATE | _____

EXCORIATE NO ONE.
*RESEARCH THIS WORD.*
*1 John 4:1 NIV*

DELIVERANCE DATE | _____

TABLEAU YOUR HOME.
*CHECK YOUR DICTIONARY.*
*1 John 4:1 ESV*

DELIVERANCE DATE | _____

BELOVED, DO NOT BELIEVE EVERY SPIRIT,
BUT TEST THE SPIRITS, WHETHER THEY ARE OF GOD.
BECAUSE MANY FALSE PROPHETS
HAVE GONE OUT INTO THE WORLD.
*1 John 4:1*

DELIVERANCE DATE | _____

THOSE WHO SEEK THE LORD SHALL
NOT LACK ANYTHING.

*Psalm 34:10*

DELIVERANCE DATE | _____

ADMIT YOUR FAULTS TO ONE ANOTHER
AND PRAY FOR EACH OTHER SO
THAT YOU MAY BE HEALED.

*James 5:16*

DELIVERANCE DATE | _____

LET GOD TEACH YOU TO HOLD YOUR TONGUE.
HE WILL CAUSE YOU TO UNDERSTAND
WHEREVER YOU ERR.

*Job 6:24*

DELIVERANCE DATE | _____

KEEP YOUR **TONGUE** FROM EVIL,
AND YOUR LIPS FROM SPEAKING DECEIT.
*Psalm 34:13*

DELIVERANCE DATE | _____

THE **TONGUE** OF THE RIGHTEOUS *IS* CHOICE SILVER.

*Proverbs 10:20*

DELIVERANCE DATE | _____

DEATH AND LIFE *ARE* IN THE POWER OF THE **TONGUE.**
*Proverbs 18:21*


DELIVERANCE DATE | _____

WHOEVER GUARDS HIS MOUTH AND **TONGUE**
KEEPS HIS SOUL FROM TROUBLES.
*Proverbs 21:23*

DELIVERANCE DATE | _____

OPEN YOUR MOUTH WITH WISDOM.
ON YOUR **TONGUE** *IS* THE LAW OF KINDNESS.

*Proverbs 31:26*

DELIVERANCE DATE | _____

PURSUE PEACE WITH ALL PEOPLE, AND HOLINESS,
WITHOUT WHICH NO ONE WILL SEE THE LORD.

*Hebrews 12:14*

DELIVERANCE DATE | _____

FIND HAPPINESS IN SPIRITUAL EXPERIENCES
RATHER THAN THINGS ONLY DREAMS ARE MADE OF.

*Romans 8:6*

DELIVERANCE DATE | _____

INVEST IN OTHERS.
*Luke 6:27-31*

DELIVERANCE DATE | _____

EXPERIENCE THE JOY OF SHARING WITH OTHER PEOPLE.

*Luke 6:38*

DELIVERANCE DATE | _____

ARE YOU TAKING UP A LOT OF ROOM?
ARE YOU MAKING ROOM FOR THOSE AROUND YOU?
*Colossians 1:20*

DELIVERANCE DATE | _____

DID YOU KNOW THAT BEING A PART OF THE ELEVEN
IS BETTER THAN BEING A TEN?
*Luke 24:8-10*

DELIVERANCE DATE | _____

Number one?
There is no number lower than that.

*James 4:10*

Deliverance Date | _____

IF YOU HESITATE YOU LOSE.

*James 4:8*

DELIVERANCE DATE | _____

You either make dust or you eat it.

*Exodus 31:3*

Deliverance Date | _____

THE VIEW FROM THE DOG SLED CHANGES
ONLY FOR THE LEAD DOG.
BE A SPIRITUAL LEADER WHEN YOU LEAD.

*2 Kings 2:6-8 KJV*

DELIVERANCE DATE | _____

You can get used to anything.
*I'm just saying.*
*Philippians 4:6 KJV*

Deliverance Date | _____

STAY IN NEUTRAL.
TRANSITIONING TO A SPIRITUAL
POSITION WILL BE EASIER.
*Matthew 6:34 KJV*

DELIVERANCE DATE | _____

RELAX.
THERE'S A SWEET SPIRIT IN THIS PLACE.
*Jude 1:24*

DELIVERANCE DATE | _____

HAVE ANOTHER BLESSED DAY.

*Romans 8:11*

DELIVERANCE DATE | _____

BE THE BLESSING YOU WANT TO RECEIVE.

*Hebrews 10:22-24*

DELIVERANCE DATE | _____

Make it your business
to be make peace and find your purpose.
*Romans 8:28 KJV*

Deliverance Date | _____

GIVE YOUR LIFE TO GOD.
HE'LL MOLD IT AND SHAPE IT
AND GIVE IT RIGHT BACK TO YOU.
*John 10:10*

DELIVERANCE DATE | _____

THAT CIRCLE OF CHAOS YOU'RE IN CAN END.
STOP, DROP AND PRAY.

*Romans 1:9*

DELIVERANCE DATE | _____

Rejoice always.
*1 Thessalonians 5:16*

Deliverance Date | _____

PRAY WITHOUT CEASING.
*1 Thessalonians 5:16*

DELIVERANCE DATE | _____

GIVE THANKS FOR EVERYTHING.

*1 Thessalonians 5:16*

DELIVERANCE DATE | _____

You are fearfully and wonderfully made.

*Psalm 139:14*

Deliverance Date | _____

GET TO KNOW THE GREAT I AM.

*Exodus 3:5*

DELIVERANCE DATE | _____

GET TO KNOW YOUR SHEPHERD.

*Psalm 23:1-6*

DELIVERANCE DATE | _____

HOPE AND FAITH.
*YOU NEED THEM BOTH.*
*Romans 5:2*

DELIVERANCE DATE | _____

GRACE AND MERCY.
*THEY ARE PATERNAL TWINS.*
*2 John 1:3*

DELIVERANCE DATE | _____

WHEN YOU'RE ONLY *LOOKING* AT YOUR PROBLEMS
YOUR PROBLEMS WILL TAKE YOU DOWN.
*Matthew 14:28-33*


DELIVERANCE DATE | _____

ONE DOOR WILL SHUT.
ANOTHER ONE WILL OPEN.
*Acts 5:22-24*

DELIVERANCE DATE | _____

(PRAY) NOW TO HIM WHO IS ABLE
TO DO **EXCEEDINGLY ABUNDANTLY**
ABOVE ALL THAT WE ASK OR THINK,
ACCORDING TO THE POWER THAT WORKS IN US.
YOU'RE WELCOME.

*Ephesians 3:20*

DELIVERANCE DATE | _____

Our blessings-delivering God
is an ever-present help.
All you have to do is show him your faith.
*Done!*
*Psalm 46:1-11 NIV*

Deliverance Date | _____

Now that you've written down your blessings, ask God to increase your faith. You are surrounded by such a huge cloud of witnesses to your life of faith. Run with endurance the race God has set before you. Keep your eyes on Jesus. He's the right man. He's the champion who initiates and perfects your faith. *Hebrews 12:1-3 NLT*

God says, "And **t**ry Me now in **t**his," Says **t**he Lord of hosts, "If I will not open for you **t**he windows of heaven and pour out for you *such* blessing **t**hat *there will* not *be* **room** **enough** **t**o receive it." *Malachi 3:10*

Dear Heart, it's all about your measure of faith. "So Jesus said to them, "Because of your unbelief; for assuredly, I say to you, if you have faith as a **mustard seed**, you will say to this mountain, 'Move from here to there,' and it will move; and nothing will be impossible for you." Matthew 17:20

Study Hebrews 11:2–12:3 below and let God see your faith too.

Through their faith, the people in days
of old earned a good reputation.
By faith we understand that the entire universe
was formed at God's command,
that what we now see did not come from
anything that can be seen.

It was by faith that Abel brought a more
acceptable offering to God than Cain did.
Abel's offering gave evidence that he was a righteous man,
and God showed his approval of his gifts.
Although Abel is long dead,
he still speaks to us by his example of faith.

It was by faith that Enoch was taken
up to heaven without dying—
"he disappeared, because God took him."
For before he was taken up, he was known
as a person who pleased God.
And it is impossible to please God without faith.
Anyone who wants to come to him
must believe that God exists
and that he rewards those who sincerely seek him.

It was by faith that Noah built a large boat
to save his family from the flood.
He obeyed God, who warned him about
things that had never happened before.
By his faith Noah condemned the rest of the world,
and he received the righteousness that comes by faith.

It was by faith that Abraham obeyed when
God called him to leave home
and go to another land that God would
give him as his inheritance.
He went without knowing where he was going.
And even when he reached the land God promised him,
he lived there by faith—for he was like
a foreigner, living in tents.
And so did Isaac and Jacob, who inherited the same promise.
Abraham was confidently looking forward
to a city with eternal foundations,
a city designed and built by God.

It was by faith that even Sarah was able to have a child,
though she was barren and was too old.
She believed that God would keep his promise.

And so a whole nation came from this one
man who was as good as dead—
a nation with so many people that,
like the stars in the sky and the sand on the seashore,
there is no way to count them.

All these people died still believing
what God had promised them.
They did not receive what was promised,
but they saw it all from a distance and welcomed it.
They agreed that they were foreigners
and nomads here on earth.
Obviously people who say such things are looking forward
to a country they can call their own.
If they had longed for the country they came from,
they could have gone back. But they
were looking for a better place,
a heavenly homeland. That is why God is
not ashamed to be called their God,
for he has prepared a city for them.

It was by faith that Abraham offered Isaac as a sacrifice
when God was testing him. Abraham,
who had received God's promises,
was ready to sacrifice his only son, Isaac,
even though God had told him,
"Isaac is the son through whom your
descendants will be counted."
Abraham reasoned that if Isaac died,
God was able to bring him back to life again.
And in a sense, Abraham did receive
his son back from the dead.

It was by faith that Isaac promised
blessings for the future to his sons,
Jacob and Esau. It was by faith that Jacob,
when he was old and dying,
blessed each of Joseph's sons and bowed in
worship as he leaned on his staff.

It was by faith that Joseph, when he was about to die,
said confidently that the people of Israel would leave Egypt.
He even commanded them to take his
bones with them when they left.

It was by faith that Moses' parents hid him
for three months when he was born.
They saw that God had given them an unusual child,
and they were not afraid to disobey the king's command.

It was by faith that Moses, when he grew up,
refused to be called the son of Pharaoh's daughter.
He chose to share the oppression of God's people
instead of enjoying the fleeting pleasures of sin.
He thought it was better to suffer for the sake of Christ
than to own the treasures of Egypt, for he
was looking ahead to his great reward.

It was by faith that Moses left the land of
Egypt, not fearing the king's anger.
He kept right on going because he kept his
eyes on the one who is invisible.
It was by faith that Moses commanded the people of Israel
to keep the Passover and to sprinkle blood on the doorposts
so that the angel of death would not kill their firstborn sons.

It was by faith that the people of Israel
went right through the Red Sea
as though they were on dry ground.
But when the Egyptians tried to follow,
they were all drowned.

It was by faith that the people of Israel
marched around Jericho
for seven days, and the walls came crashing down.

It was by faith that Rahab the prostitute was not destroyed
with the people in her city who refused to obey God.
For she had given a friendly welcome to the spies.

How much more do I need to say?
It would take too long to recount the stories of the faith of
Gideon, Barak, Samson, Jephthah, David,
Samuel, and all the prophets.
By faith these people overthrew kingdoms, ruled with justice,
and received what God had promised them.
They shut the mouths of lions, quenched the flames of fire,
and escaped death by the edge of the sword.
Their weakness was turned to strength.
They became strong in battle and put whole armies to flight.
Women received their loved ones back again from death.
But others were tortured, refusing to turn
from God in order to be set free.
They placed their hope in a better life after the resurrection.
Some were jeered at, and their backs
were cut open with whips.
Others were chained in prisons. Some died by stoning,
some were sawed in half, and others
were killed with the sword.

Some went about wearing skins of sheep and goats,
destitute and oppressed and mistreated.
They were too good for this world, wandering
over deserts and mountains,
hiding in caves and holes in the ground.
All these people earned a good reputation
because of their faith,
yet none of them received all that God had promised.
For God had something better in mind for us,
so that they would not reach perfection without us.

Therefore, since we are surrounded by such a
huge crowd of witnesses to the life of faith,
let us strip off every weight that slows us down,
especially the sin that so easily trips us up.
And let us run with endurance the
race God has set before us.
We do this by keeping our eyes on Jesus,
the champion who initiates and perfects our faith.
Because of the joy awaiting him, he endured the cross,
disregarding its shame.
Now he is seated in the place of honor beside God's throne.
Think of all the hostility he endured from sinful people;
then you won't become weary and give up.

Begin another *Deliverance Diary*. It's all about your faith.

## About the Author

# It's about giving flowers...

Since forever, we've heard our families and friends say, "Give me my flowers while I'm living." It is these words that inspired the creation of a line of greeting cards, books, and Christian products we call a few flowers™.

We offer elegant, limited edition, handmade cards written to encourage, inspire, strengthen, and assure.

There are Christian books that will help you heal and products that will delight you in a mighty way.
*Look us over at afewflowers.com.*

*Dearly Beloved,*
*The flowers are in the words.*
*The words are the flowers*
*that are being given to you.*
*My goal is to give you*
*your flowers while you're living.*
*Therefore, here are*
*a few flowers™ from me to you.*

E. J. H. MOFFETT is a member of the Church of Christ at East Side in Austin, Texas.

She has an associate's degree in arts from Austin Community College, a bachelor's degree in journalism from the University of Texas at Austin, and a master of liberal arts degree from St. Edwards University, all located in Austin, Texas. She has owned her Christian greeting card company, a few flowers™, since 2005. She also creates church forms and documents, and handles communications for her church.

Erma has been prayerfully and compassionately studying the Word of God for more than thirty-five years. Grounded in her study and understanding of the scriptures, her elegant, limited-edition Christian greeting cards continue to encourage, inspire, strengthen, and assure her customers. She is the author of the book Begin Your Healing…Write to God.

This Deliverance Diary is just another opportunity for you to write to God. It is Erma's goal to give flowers to those she meets on her journey. This Deliverance Diary gives you a way to give flowers too.

It's all about your measure of faith – how much you have; the amount of you allow God to see; how it shows up in your troubles; and how you hold fast to it in your storms. It is her prayer that you will write down every blessing.

*So then faith comes
by hearing,
and hearing
by the word of God.*
Romans 10:17

# The Scripture Passage That Began My Journey

But seek first the kingdom of God and his righteousness,
and all these things will be added to you.
"Therefore do not be anxious about tomorrow,
for tomorrow will be anxious for itself.
Sufficient for the day is its own trouble.
*Meditate on Matthew 6:33, 34 ESV.*

But seek first his kingdom and his righteousness,
and all these things will be given to you as well.
Therefore do not worry about tomorrow,
for tomorrow will worry about itself.
Each day has enough trouble of its own.
*Meditate on Matthew 6:33, 34 NIV.*

But seek ye first the kingdom of God, and his righteousness;
and all these things shall be added unto you.
Take therefore no thought for the morrow:
or the morrow shall take thought for the things of itself.
Sufficient unto the day is the evil thereof.
*Meditate on Matthew 6:33, 34 KJV.*

*Everything you need
is in the church.*